Spring Serenade

by Sherry Bonnice

A Harmony of Wool & Cotton
Appliqué Projects

CHITRA PUBLICATIONS

Your Best Value in Quilting!
www.QuiltTownUSA.com

Chitra Publications
2 Public Avenue
Montrose, Pennsylvania 18801-1220

First Printing: 2002
Library of Congress Cataloging-in-Publication Data

Bonnice, Sherry, 1956-
Spring serenade : a harmony of wool and appliqué projects / by Sherry Bonnice.
        p. ; cm.
    ISBN 1-885588-43-7
        1. Appliqué—Patterns. 2. Patchwork—Patterns. 3. Wool quilts.   I. Title.
    TT779.B653 2002
    746.44'5—dc21

                                        2002018806

Edited by: Deborah Hearn
Design and Illustrations: K.A. Steele and Jennifer Marie Oakes
Photography: Van Zandbergen Photography, Brackney, Pennsylvania
Cover Photography: Guy Cali Photography, Inc. Clarks Summit, Pennsylvania

Our Mission Statement

We publish quality quilting magazines and books that recognize, promote, and inspire self-expression. We are dedicated to serving our customers with respect, kindness, and efficiency.

Dear Readers,

A friend once shared with me a creed I now live by: "I don't take the time to make anything unless I absolutely love it." Making decisions based on this precept has changed my creative life! Before hearing it, I arbitrarily tried patterns or crafts to see if I liked a technique or a look. Afterward, everything I considered making went through a critical filtering system that included the "BIG" question. "Do I absolutely love it?"

So, when I began designing projects for this book, it was easy to choose motifs I love: bunnies, flowers, sheep, birds, and ducks. As I used the designs in the quilts, pillows, and wallhangings, my excitement continued to grow. What a joy it is to share these designs in the beautiful and creative medium of appliqué.

My decision to include wool appliqué was also an easy one. After my first experience making a wool appliqué pin cushion, I was hooked. I hope that, while trying these easy-to-do wool designs, you'll discover you love wool appliqué too. You might even love wool in bright colors that make you smile.

The projects in this book include wallhangings, pillows, and quilts in colors that make them sing. When selecting fabric for the designs you wish to make, choose happy spring colors that shout new life in a wonderful serenade. Whether you decide to stitch your projects using needleturn appliqué or a blanket stitch, I hope you'll enjoy creating something you absolutely love.

Sherry

P.S. A special thanks to Barbara Tibus for machine quilting "Pinwheels for Posie" and "Spring Song." Barb owns and operates Quilts to Treasure a custom machine quilting business. To contact her, write to: Barabara R. Tibus, Quilts to Treasure, 886 Shoemaker Avenue, West Wyoming, PA 18644. (570)693-0507 email:btibus@Adelphia.net

Contents

Say Hi to Hyacinths!

Enjoy early spring flowers all year long with this garden mini quilt and pillow.

Purple and green look so pretty together, but Hyacinths are beautiful in pink and white too. Choose your favorite color, and stitch "Say Hi to Hyacinths!" to celebrate color, spring, and beautiful flowers. Try the smaller version to make a "Little Hyacinth Pillow" to match.

Materials for Quilt

FINISHED SIZE: 15" square
BLOCK SIZE: 2" square

* Purple, gold, and dark green solid, each at least 8" square, for the flowers
* 1/4 yard muslin
* Fat quarter (18" x 22") purple stripe
* 3/8 yard green print
* 17" square of backing fabric
* 17" square of thin batting
* Freezer paper

Materials for Pillow

FINISHED SIZE: 5 1/2" square
BLOCK SIZE: 1 3/4" square

* Purple, gold, and dark green solid, each at least 1 1/2" square, for the flowers
* Piece of muslin at least 4" square
* Piece of purple stripe at least 8" square
* Piece of green print at least 10" square
* 8" square of backing fabric
* 8" square of thin batting
* Fiberfill
* Freezer Paper

CUTTING

Appliqué pattern pieces are full size and do not include a turn-under allowance. Cut 13 each of pieces #1 through #3 using appropriately colored fabrics. All other dimensions include a 1/4" seam allowance.

- Cut 26: 1 7/8" squares, muslin
- Cut 1: 1 1/2" x 44" strip, muslin
- Cut 26: 1 7/8" squares, purple stripe
- Cut 2: 1" x 12" strips, purple stripe, with the stripes running lengthwise, for the inner border
- Cut 2: 1" x 13" strips, purple stripe, with the stripes running widthwise, for the inner border
- Cut 4: 2 1/4" x 16" strips, green print, for the outer border
- Cut 2: 1 1/4" x 44" strips, green print, for the binding
- Cut 1: 1 1/2" x 44" strip, green print

DIRECTIONS

- Place a 1 7/8" purple stripe square, wrong side up, on the work surface with the stripes running vertically. Draw a diagonal line from the upper right corner to the lower left corner. Make 13. In the same manner, draw a diagonal line from the upper left corner to the lower right corner on the remaining 1 7/8" purple stripe squares.
- Lay a marked 1 7/8" purple stripe square on a 1 7/8" muslin square, right sides together. Stitch 1/4" away from the drawn line on both sides. Make 26.
- Cut the squares on the drawn lines to yield 52 pieced squares. Press the seam allowances toward the purple stripe.
- Lay out 4 pieced squares, as shown.

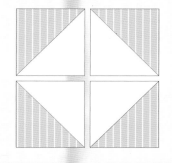

Stitch the units into rows and join the rows to make a pieced block. Make 13.
- Position pieces #1 and #2 on a pieced block, referring to the Placement Diagram. Pin and needleturn appliqué piece #1 in place.
- In the same manner, appliqué piece #2 in place.
- Center piece #3 on the upper portion of piece #1. Pin and appliqué it in place. Make 13.
- Stitch the 1 1/2" x 44" muslin strip and the 1 1/2" x 44" green print strip, right sides together along their length. Press the seam allowance toward the green print. From this pieced panel, cut twenty-four 1 1/2" slices.

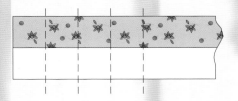

- Stitch two slices into a Four Patch, as shown. Make 12.

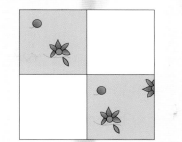

- Referring to the photo, lay out the appliqué blocks and the Four Patch blocks in 5 rows of 5. Stitch the blocks into rows and join the rows.
- Measure the length of the quilt. Trim the 1" x 12" lengthwise purple stripe strips to that measurement. Stitch them to the sides of the quilt.
- Measure the width of the quilt including the borders. Trim the 1" x 13" widthwise purple stripe strips to that measurement. Stitch them to the top and bottom

of the quilt.
- In the same manner, trim two 2 1/4" x 16" green print strips to fit the quilt"s length and stitch them to the sides of the quilt.
- Trim the remaining 2 1/4" x 16" green print strips to fit the quilt's width and stitch them to the top and bottom of the quilt.
- Finish the quilt according to *Stitching Tips* (page 31), using the 1 1/4" x 44" green print strips for the binding.

Pattern Pieces & Placement Diagram

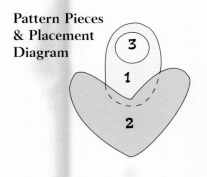

Little Hyacinth Pillow

CUTTING

Cut 2 each of pieces #1 through #3 using appropriately colored fabrics.
- Cut 4: 1 3/4" squares, muslin
- Cut 4: 1 3/8" squares, muslin
- Cut 4: 1 3/4" squares, purple stripe
- Cut 1: 1 1/2" x 5" strip, purple stripe, with the stripes running widthwise, for the border
- Cut 1: 1 1/2" x 6" strip, purple stripe, with the stripes running widthwise, for the border
- Cut 1: 6" square, green print, for the pillow back
- Cut 4: 1 3/8" squares, green print
- Cut 1: 1 1/2" x 5" strip, green print, for the border
- Cut 1: 1 1/2" x 4" strip, green print, for the border

DIRECTIONS
- Following directions for the quilt, use the 1 3/4" purple stripe squares and 1 3/4" muslin squares to make 2 appliqué blocks. Appliqué the appropriate pieces on each block.
- Following directions for the quilt, use

the 1 3/8" muslin squares and 1 3/8" green print squares to make 2 Four Patch blocks.
- Lay out the appliqué blocks and Four Patch blocks in 2 rows of 2. Stitch them into rows and join the rows to make the pillow front.

Pattern Pieces & Placement Diagram

- Stitch the 1 1/2" x 4" green print strip to the left side of the pillow front.
- Working clockwise, stitch the 1 1/2" x 5" purple stripe strip to the top. Stitch the 1 1/2" x 5" green print strip to the right side, and the 1 1/2" x 6" purple stripe strip to the bottom of the pillow front.
- Layer the 8" square of muslin with the 8" square of thin batting and the pillow top. Quilt the top as desired.
- Trim the top to 6" square.
- With right sides together, stitch the pillow top to the pillow back leaving an opening to turn and stuff. Turn the pillow right side out through the opening and stuff with fiberfill to desired firmness. Slipstitch the opening closed.

Daffodil Bunnies

Soft bunnies & pretty flowers make sweet dreams for a special baby.

Materials for Quilt

FINISHED SIZE: 31" x 47"
BLOCK SIZE: 8 1/4" square

- Fat quarter (18" x 22") medium brown print for the bunnies
- Fat eighth (11" x 18") dark brown print for the bunny legs and ears
- Light green print, at least 12" square, for the stems
- Fat eighth of medium green print for the leaves
- Gold print, at least 12" square, for the petals
- 1 3/4 yards blue check
- 5/8 yard yellow print
- 1 1/2 yards of backing fabric
- 35" x 51" piece of thin batting
- Dark gray embroidery floss
- Freezer Paper

Materials for Pillow (pg. 8)

FINISHED SIZE: 14" square

- Medium brown print, at least 7" square, for the bunny
- Dark brown print, at least 5" square, for the bunny leg and ear
- Light green print, at least 2" x 8", for the stem
- Medium green print, at least 3" x 7", for the leaves
- Gold print, at least 4" square, for the petals
- 1/2 yard blue check
- 1/4 yard yellow print
- 16" square of thin batting
- 16" square muslin
- 14" pillow form
- Dark gray embroidery floss
- Freezer Paper

CUTTING

Appliqué patterns are full size and do not include a turn-under allowance. Cut 7 each of pieces #1 through #10 using appropriately colored fabrics. All other dimensions include a 1/4" seam allowance.

- Cut 7: 9 1/4" squares, blue check
- Cut 32: 3 1/4" squares, blue check
- Cut 8: 1 7/8" x 44" strips, blue check
- Cut 5: 2 1/2" x 44" strips, blue check, for the binding
- Cut 8: 3 1/4" squares, yellow print
- Cut 8: 1 7/8" x 44" strips, yellow print, for the border

DIRECTIONS

- Position pieces #1, #2, #3, and #9 on a 9 1/4" blue check square, referring to the Placement Diagram. Rearrange the pieces until you like the placement, keeping the outer edges at least 3/4" from the raw edges of the square.
- Remove pieces #3 and #9. Pin and needleturn appliqué pieces #1 and #2 in place.
- Working in numerical order and referring to the Placement Diagram, needleturn appliqué pieces #3 through #10 on the blue check square.
- Using 2 strands of the dark gray

Make any baby's room special with this sweet ensemble of curious bunnies enjoying spring's flowers and beauty. "Daffodil Bunnies" will bring a smile to mom too. Note: Pillows and buttons (or plastic eyes) can be dangerous for babies and young children. Use them for decorative purposes only until your child is older.

embroidery floss embroider an eye on the bunny using a satin stitch. Refer to page 28 for a stitch diagram.

- Trim the appliquéd square to 8 3/4" keeping the design centered. Make 7.
- Stitch a 1 7/8" x 44" blue check strip to a 1 7/8" x 44" yellow print strip, right sides together along their length, to make a long pieced strip. Press the seam allowance toward the blue check. Make 8.
- Cut seventy-two 1 7/8" slices from 4 of the long pieced strips. NOTE: *Cut 21 slices from each of 3 strips and 9 from the fourth strip. Save the leftover portion from the fourth strip for the pillow.*

- Stitch two slices into a Four Patch, as shown. Make 36.
- Lay out 4 Four Patches, four 3 1/4" blue check squares, and one 3 1/4"

Pattern Pieces & Placement Diagram

yellow print square in 3 rows of 3 as shown. Stitch them into rows and join the rows to form a Nine Patch block. Make 8.

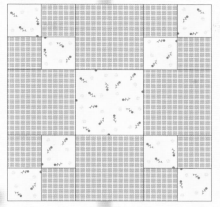

- Referring to the photo, lay out the appliqué blocks and the Nine Patch blocks in 5 rows of 3. Stitch the blocks into rows and join the rows.

- Measure the width of the quilt. Trim 2 of the long pieced strips to that measurement. Stitch a Four Patch block to each end of the trimmed strips. Make 2. Set them aside.

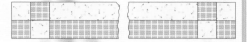

- Measure the length of the quilt. Trim the remaining long pieced strips to that measurement. Stitch them to the sides of the quilt, placing the yellow print strips against the quilt center.
- Stitch the pieced strips you set aside to the top and bottom of the quilt.
- Finish the quilt according to *Stitching Tips* (page 31), using the 2 1/2" x 44" blue check strips for the binding.

Daffodil Bunny Pillow

CUTTING

Appliqué patterns (page 7) are full size and do not include a turn-under allowance. Cut pieces #1 through #10 using appropriately colored fabrics. All other dimensions include a 1/4" seam allowance.

- Cut 2: 14 1/2" squares, blue check, for the pillow back
- Cut 1: 10" square, blue check
- Cut 2: 1 7/8" x 28" strips, blue check, for the border
- Cut 2: 1 7/8" x 28" strips, yellow print, for the border

DIRECTIONS

- In the same manner as for the quilt, needleturn appliqué the bunny and daffodil to the 10" blue check square.
- Using 2 strands of the dark gray embroidery floss, embroider an eye on the bunny using a satin stitch. Refer to page 28 for a stitch diagram.
- Trim the appliquéd square to 9", keeping the design centered.
- Stitch a 1 7/8" x 28" blue check strip and a 1 7/8" x 28" yellow print strip, right sides together along their length. Press the seam allowance toward the blue check. Make 2.
- Cut four 1 7/8" slices and two 9" strips from each pieced strip.
- Stitch two 1 7/8" slices into a Four Patch. Make 4.
- Stitch a 9" pieced strip to each of two opposite sides of the pillow top.
- Stitch a Four Patch unit to each end of the remaining 9" pieced strips.
- Stitch them to the remaining sides of the pillow top.
- Layer the pillow top with the 15" square of thin batting and the 15" square of muslin. Quilt as desired.
- To make the pillow back, press under a 1/4" hem on one edge of each 14 1/2" blue check square. Stitch the hem in place.
- Fold under 4" along the same edge and press but do not stitch.
- Lay the two pillow back pieces, right side up, with pressed edges overlapping. Adjust the overlap until the pillow measures 14 1/2" square. Baste in the top and bottom seam allowances to secure the overlapped edges.
- Pin the pillow front to the back, right sides together. Stitch around all sides with a 1/4" seam allowance.
- Turn the pillow right side out through the overlapped opening and insert the pillow.

Daffodil Bunny Toy

Make a stuffed bunny to add to your spring collection.

*Make "**Daffodil**" in your favorite cotton prints. Try one in wool too. This easy-to-make stuffed bunny will be multiplying like, well, like rabbits, in no time at all.*

Materials for Bunny

FINISHED SIZE: 7" high x 9" long
- Fat quarter (18" x 22") primary fabric
- Fat eighth (11" x 18") contrasting fabric
- 1 pound fiberfill
- 2 buttons or plastic eyes
- Upholstery thread to match the primary fabric
- Long needle for attaching the legs

CUTTING

Dimensions include a 1/4" seam allowance.

- Cut 1: underside piece and one piece reversed, primary fabric
- Cut 1: head/body piece and one piece reversed, primary fabric
- Cut 1: leg piece and one piece reversed primary fabric
- Cut 2: ear pieces, primary fabric
- Cut 1: leg piece and one piece reversed contrasting fabric
- Cut 2: ear pieces, contrasting fabric

DIRECTIONS

- Stitch the underside pieces, right sides together, along the center seam.
- Stitch the head/body pieces, right sides together, along the head and back edge from point A to point B.
- Pin the underside piece to the head/body piece from point A, around the leg area, to point B. Stitch, leaving an opening on one side for turning, as marked on the pattern.
- Clip the seam allowances along the curves, being careful not to clip the thread, and turn the body right side out. Stuff firmly with fiberfill.
- Stitch the opening closed using a ladder stitch.
- Stitch one primary fabric leg and one contrasting leg right sides together, leaving an opening for turning, as marked on the pattern.
- Clip the seam allowances along the curves and turn the leg right side out. Stuff firmly.

- Stitch the opening closed using a ladder stitch. Make 2.
- Stitch one primary fabric ear and one contrasting ear right sides together, leaving the bottom open for turning.
- Clip the seam allowances along the curves and turn the ears right side out.
- Stitch the bottom closed using an overedge stitch and gathering tightly. Make 2.
- Pin the ears in position on the head. Stitch them in place using an overedge stitch.
- Stitch the legs to the sides of the body in the following manner, using upholstery thread. Make a 1/2" stitch on the inside of the right leg.

Continue through the body and make a 1/2" stitch on the inside of the left leg. Continue back through the body, as shown, pulling the ends of the thread to tighten the legs. Knot the ends securely.

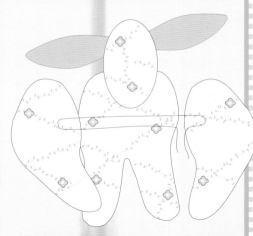

- Stitch the eyes in place in the same manner.

Underside

Center Seam ↓

Leg

Opening

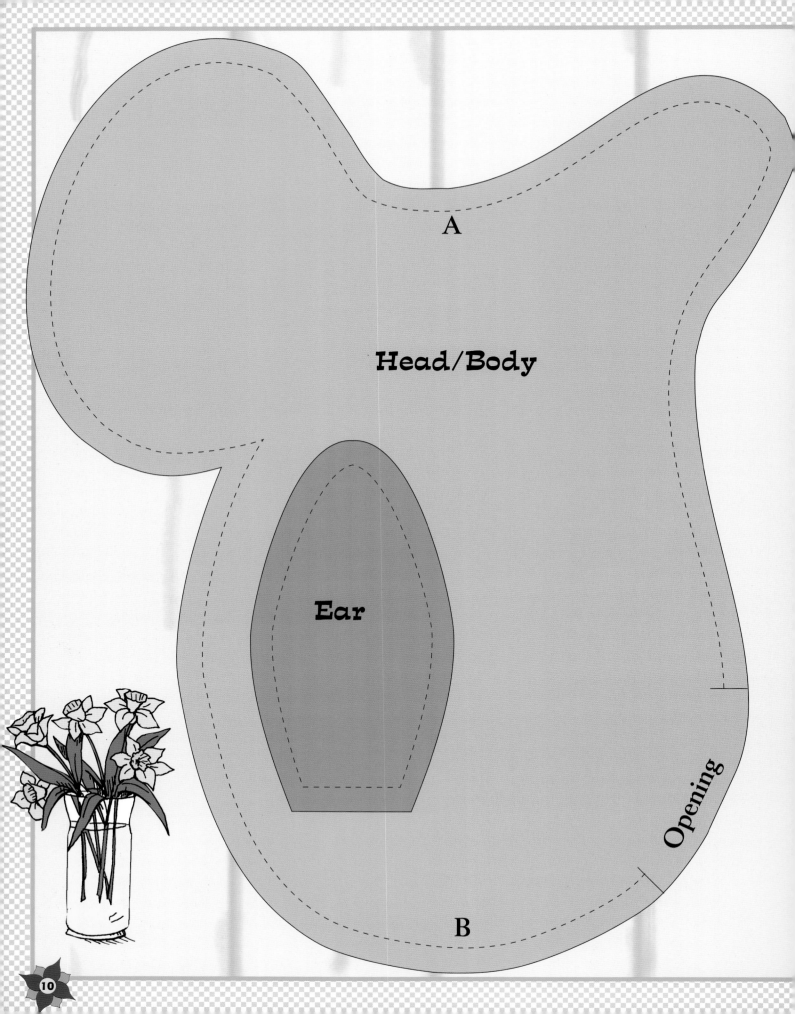

A

Head/Body

Ear

Opening

B

A Taste of Honey

Make a fun quilt & pillow that are almost sweet enough to eat.

Keeping his eyes on the prize, this bear is looking forward to "A Taste of Honey." The wallhanging features needle-turn appliqué in bright spring colors.

"A Taste of Honey Pillow" offers a quick project made with textured wools that are perfect for this fuzzy fellow.

Materials for Quilt

FINISHED SIZE: 15" square
- Assorted brown prints for the bear, limb, and beehive
- Assorted greens for the leaves
- Gold scrap for the bees
- 3/8 yard blue print
- 1/8 yard yellow print
- 17" square of backing fabric
- 17" square of thin batting
- Embroidery floss in white and black

Materials for Pillow (pg. 13)

FINISHED SIZE: 16" square
Yardage for flannel is based on 44" fabric with a useable width of 42".
- 1/2 yard green check flannel
- Assorted brown wool scraps for the bear, limb, and beehive
- Assorted green wool scraps for the leaves
- Gold wool scraps for the bees
- Blue wool at least 9 1/2" square for the background
- 16" pillow form
- Wool embroidery thread, embroidery floss, or pearl cotton in colors to match or contrast with the wool scraps
- Embroidery floss in off-white, white, and black

CUTTING

Appliqué patterns (page 12) are full size and do not include a turn-under allowance. Cut out 1 each of pieces #1 through #19 using appropriately colored fabrics. All other dimensions include a 1/4" seam allowance.
- Cut 1: 10" square, blue print
- Cut 4: 1 1/4" x 9 1/2" strips, blue print
- Cut 2: 1 1/4" x 44" strips, blue print, for the binding
- Cut 4: 1 1/4" x 2 1/2" rectangles, blue print
- Cut 4: 1 1/4" x 1 3/4" rectangles, blue print
- Cut 4: 2 1/2" x 11" strips, yellow print
- Cut 4: 1 3/4" squares, yellow print
- Cut 4: 1 1/4" squares, yellow print

DIRECTIONS

- Position pieces #1 and #2 on the 10" blue print square, referring to the Placement Diagram. Pin and needle-turn appliqué them in place.
- Working in numerical order and referring to the photo for placement, needleturn appliqué pieces #3 through #19 on the blue print square.
- Trim the appliquéd blue print square to 9 1/2".
- Stitch two 1 1/4" x 9 1/2" blue print strips to the sides of the quilt.
- Stitch a 1 1/4" x 9 1/2" blue print strip between two 1 1/4" yellow print squares. Make 2. Stitch them to the top and bottom of the quilt.
- In the same manner, stitch two 2 1/2" x 11" yellow print strips to the sides of the quilt.
- Stitch a 1 1/4" x 1 3/4" blue print rectangle to a 1 3/4" yellow print square.
- Stitch a 1 1/4" x 2 1/2" blue print rectangle to this pieced rectangle to make a pieced square. Make 4.

- Stitch a pieced square to each end of the remaining 2 1/2" x 11" yellow print strips.
- Stitch them to the remaining sides of the quilt.

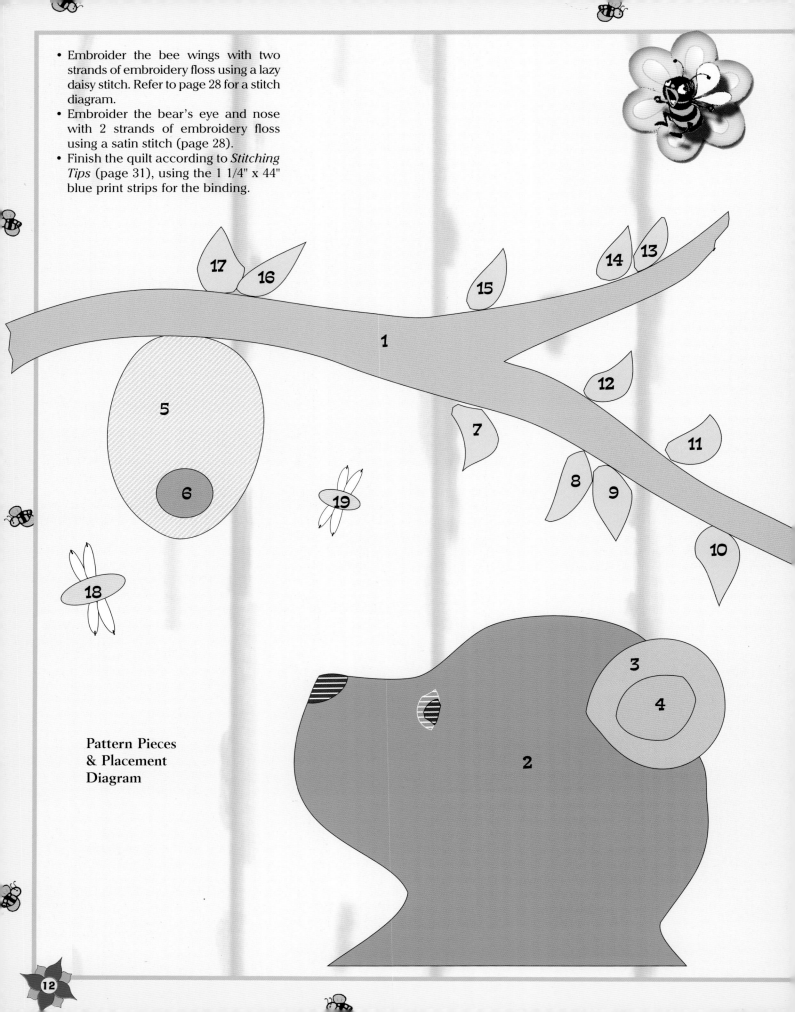

- Embroider the bee wings with two strands of embroidery floss using a lazy daisy stitch. Refer to page 28 for a stitch diagram.
- Embroider the bear's eye and nose with 2 strands of embroidery floss using a satin stitch (page 28).
- Finish the quilt according to *Stitching Tips* (page 31), using the 1 1/4" x 44" blue print strips for the binding.

**Pattern Pieces
& Placement
Diagram**

A Taste of Honey Pillow

CUTTING

Pattern pieces are full size and do not need a turn-under allowance. Cut pieces #1 through #19 using appropriately colored wool. All other dimensions include a 1/4" seam allowance.

- Cut 1: 9 1/2" square, blue wool
- Cut 3: 17" squares, green check flannel, one for the pillow front and two for the pillow back

DIRECTIONS

- Position pieces #1 and #2 on the 9 1/2" blue wool square, referring to the Placement Diagram on the facing page. Pin and blanket stitch them in place. Refer to page 28 for a stitch diagram.
- Working in numerical order and referring to the photo for placement, blanket stitch pieces #3 through #19 on the blue wool square.
- Referring to the photo, embroider the bee wings with two strands of embroidery floss using a lazy daisy stitch.
- Embroider the eye and nose with two strands of embroidery floss using a satin stitch.
- Center and blanket stitch the blue wool square onto one of the 17" green flannel squares.
- To make the pillow back, press under a 1/4" hem on one edge of each remaining green flannel square. Stitch the hem in place.
- Fold under 4" along the same edge and press but do not stitch.
- Lay the two pillow back pieces, right side up, with pressed edges overlapping. Adjust them until the pillow back measures 17" square. Baste in the top and bottom seam allowances to secure the overlapped edges.
- Pin the pillow front to the back, right sides together. Stitch around all sides with a 1/4" seam allowance.
- Turn the pillow right side out through the overlapped opening and insert the pillow form.

Pinwheels for Posie

Anticipate a perfect spring day with these cheerful colors, sheep, & pinwheels.

Materials for Quilt (pg. 15)

FINISHED SIZE: 83" square
BLOCK SIZE: 12" square

- 5 1/4 yards yellow print
- 2 3/8 yards pink print
- 2 yards green print
- Fat quarter (18" x 22") white print
- Scrap of black print at least 10" square
- 7 1/8 yards of backing fabric
- 87" square of batting

CUTTING

Appliqué patterns (page 14) are full size and do not include a turn-under allowance. All other dimensions include a 1/4" seam allowance. Cut the lengthwise green print strips before cutting other pieces from the same yardage.

For the Pinwheel blocks:
- Cut 13: 8" squares, yellow print
- Cut 52: 3 1/2" squares, yellow print
- Cut 26: 3 7/8" squares, yellow print
- Cut 13: 8" squares, pink print
- Cut 26: 3 7/8" squares, green print

For the appliqués:
- Cut 8: sheep bodies, white print
- Cut 16: feet, black print
- Cut 8: faces, black print
- Cut 8: ears, black print

Also:
- Cut 9: 2 1/2" x 44" strips, yellow print, for the binding
- Cut 2: 18 1/4" squares, yellow print, then cut them in quarters diagonally to yield 8 setting triangles
- Cut 2: 9 3/8" squares, yellow print, then cut them in half diagonally to yield 4 corner triangles
- Cut 16: 7 1/4" x 10 1/2" rectangles, yellow print
- Cut 4: 7 1/4" squares, yellow print
- Cut 8: 4" x 7 1/4" strips, yellow print
- Cut 36: 4" x 12 1/2" strips, pink print
- Cut 12: 2 1/2" x 7 1/4" strips, pink print, for the sashing
- Cut 4: 2" x 70" lengthwise strips, green print
- Cut 12: 4" squares, green print
- Cut 3: 6 1/4" squares, green print, then cut them in quarters diagonally to yield 12 sashing triangles
- Cut 8: 4 1/2" x 7 1/4" strips, green print
- Cut 8: 3 1/2" x 7 1/4" strips, green print

PREPARATION

- Draw diagonal lines from corner to corner on the wrong side of each 8" yellow print square. Draw horizontal and vertical lines through the centers.

- Draw a diagonal line from corner to corner on the wrong side of each 3 7/8" yellow print square.

DIRECTIONS

- Referring to the photo (page 15), place 2 legs, a face, and a sheep body on a yellow print setting triangle. Arrange the pieces until you like the placement, keeping the outer edges at least 3/4" from the raw edges of the triangle.
- Remove the sheep body and pin the feet and face in place. Needleturn appliqué them to the setting triangle.

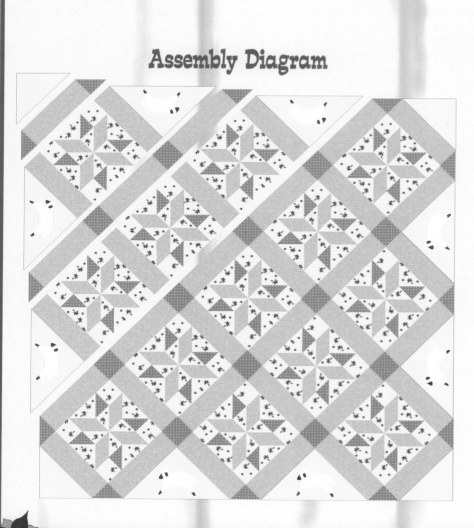

Assembly Diagram

- Pin and appliqué the sheep's body in place.
- Pin and appliqué an ear in place. Make 8.

For the Pinwheel blocks:

- Lay a marked 8" yellow print square on an 8" pink print square, right sides together. Stitch 1/4" away from the diagonal lines on both sides. Make 13.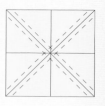
- Cut the squares on the drawn lines to yield 104 yellow/pink pieced squares. Press the seam allowances toward the pink print. Trim the squares to 3 1/2".
- Lay a marked 3 7/8" yellow print square on a 3 7/8" green print square, right sides together, and stitch 1/4" away from the marked line on both sides. Make 26.

To express the joy that my daughter Lindsey and I felt while raising our lamb, Posie, I appliquéd likenesses of her romping around brightly colored Pinwheels in the quilt appropriately named "Pinwheels for Posie."

- Cut the squares on the drawn lines to yield 52 yellow/green pieced squares. Press the seam allowances toward the green print.
- Lay out 8 yellow/pink pieced squares, 4 yellow/green pieced squares, and four 3 1/2" yellow print squares, as shown. Stitch the squares into rows. Join the rows to make a Pinwheel block. Make 13.

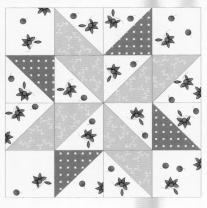

ASSEMBLY

- Referring to the Assembly Diagram, lay out the Pinwheel blocks, the 4" x 12 1/2" pink print strips, the 4" green print squares, the green print sashing triangles, the yellow print setting triangles, and the yellow print corner triangles.
- Stitch them into diagonal rows. Join the rows.
- Measure the width of the quilt. Trim 2 of the 2" x 70" green print strips to that measurement. Stitch them to opposite sides of the quilt.
- Measure the length of the quilt, including the borders. Trim the remaining 2" x 70" green print strips to that measurement. Stitch them to the remaining sides of the quilt.

For the Pieced border:

- Referring to the photo, lay out four 7 1/4" x 10 1/2" yellow print rectangles, two 4" x 7 1/4" yellow print strips, three 2 1/2" x 7 1/4" pink print strips, two 4 1/2" x 7 1/4" green print strips, and two 3 1/2" x 7 1/4" green print strips.
- Stitch them together along their 7 1/2" sides to make a pieced border. Make 4.
- Measure the width of the quilt. Trim the pieced borders to that measurement, being sure to cut an even amount from each end. Stitch trimmed borders to 2 opposite sides of the quilt.
- Stitch a 7 1/4" yellow print square to each end of the remaining pieced borders to make a long pieced border. Make 2.
- Stitch them to the remaining sides of the quilt.
- Finish the quilt as described in *Stitching Tips* (page 31), using the 2 1/2" x 44" yellow print strips for the binding.

Spring Song

Create your own songbird chorus.

Materials for Quilt

FINISHED SIZE: 52" x 66"
BLOCK SIZE: 6" x 8"

* 7/8 yard off-white print
* 1 1/4 yards pink print
* 1 yard yellow print
* 1 5/8 yards turquoise print
* Fat quarter (18" x 22") dark turquoise print
* Fat eighth (11" x 18") cream print
* Fat eighth rose print
* 3 yards backing fabric
* 56" x 70" piece of batting
* Fusible web
* Embroidery floss or sewing thread in assorted colors to match or contrast with the fabrics

CUTTING

Appliqué patterns are full size and do not include a turn-under allowance. Trace the patterns (opposite) on the paper side of the fusible web. Rough cut around the traced shapes. Fuse the shapes to the wrong side of appropriate fabrics and cut them out on the lines. All other dimensions include a 1/4" seam allowance. Cut lengthwise turquoise print strips before cutting other pieces from the same yardage.

For the appliqués:

* Cut 13: bodies, dark turquoise print
* Cut 13: breasts, rose print
* Cut 13: wings, cream print

Also:

* Cut 13: 7" x 9" rectangles, off-white print
* Cut 15: 2 1/2" x 44" strips, pink print
* Cut 12: 2 1/2" x 44" strips, yellow print
* Cut 5: 2 1/2" x 54" lengthwise strips, turquoise print, for the binding
* Cut 6: 2" x 51 1/2" lengthwise strips, turquoise print, for the sashing
* Cut 42: 2" x 8 1/2" strips, turquoise print, for the sashing

DIRECTIONS

* Referring to the photo, place a body, breast, and wing on a 7" x 9" off-white rectangle. Arrange the pieces until you like the placement, keeping the outer edges at least 3/4" from the raw edges fo the rectangle adjusting as necessary to keep it centered.
* Fuse the pieces in place.
* Embroider a blanket stitch around the appliqué pieces using 2 strands of matching or contrasting embroidery floss. Refer to page 28 for a stitch dia-

*Imagine birds singing a beautiful "**Spring Song**" as you piece and appliqué this simple but appealing design. This happy, fun quilt goes together quickly, especially if you blanket stitch using your sewing machine.*

gram. *(If stitching by machine, use sewing weight thread and the blanket stitch setting on your sewing machine.)*

- Stitch bird legs and an open beak with a stem stitch, using 2 strands of tan embroidery floss. Refer to page 28 for a stitch diagram.
- Trim the block to 6 1/2" x 8 1/2", keeping the design centered. Make 13.
- Stitch a 2 1/2" x 44" yellow print strip between two 2 1/2" x 44" pink print strips, along their length. Press the seam allowances toward the pink print strips. Make 6.
- Cut four 8 1/2" slices from each pieced

strip to make the pink Rail Fence blocks.
- In the same manner, stitch a 2 1/2" x 44" pink print strip between two 2 1/2" x 44" yellow print strips. Press the seam allowances toward the pink print strip. Make 3.
- Cut four 8 1/2" slices from each pieced strip to make the yellow Rail Fence blocks.
- Referring to the photo, lay out the 13 appliqué bird blocks, the 24 pink Rail Fence blocks, the 12 yellow Rail Fence blocks, and the 2" x 8 1/2" turquoise print sashing strips. Stitch them into rows.
- Lay out the rows with the 51 1/2" turquoise sashing strips between them. Join the block rows and the sashing strips.
- Finish the quilt as described in *Stitching Tips* (page 31), using the 2 1/2" x 54" turquoise print strips for the binding.

New Arrivals

Stitch a sampling of spring's wonders.

When ducks build nests around the pond near my home, each day reveals signs of new life. Ducklings that appear to be sweet puffs of feathers swimming, dipping, and diving as they follow mom are one of Spring's best gifts. *"New Arrivals"* is a birth announcement in fabric.

Materials for Quilt

FINISHED SIZE: 22 1/2" x 28 1/2"
- 1 yard mottled blue
- 10" square of off-white print for the duck's body
- Assorted print, plaid, and/or stripe scraps for the remaining appliqués, including oranges, browns, greens, yellows, off-whites, and golds
- 1/4 yard yellow for the sashing and the binding
- 1/8 yard lavender for the sashing
- 25" x 31" piece of backing fabric
- 25" x 31" piece of thin batting

CUTTING

Appliqué patterns are full size and do not include a turn-under allowance. Cut the pieces for each block using appropriately colored fabrics. Keep the pieces for each block separate. The inside of the nest (piece #1) for the Nest block is cut as one piece
- Cut 1: 10 1/2" x 16 1/2" rectangle, mottled blue
- Cut 2: 6 1/2" x 16 1/2" rectangles, mottled blue
- Cut 2: 6 1/2" x 10 1/2" rectangles, mottled blue
- Cut 4: 6 1/2" squares, mottled blue
- Cut 4: 1 1/2" x 6" strips, yellow
- Cut 2: 1 1/2" x 16" strips, yellow
- Cut 3: 1 1/4" x 44" strips, yellow, for the binding
- Cut 2: 1 1/2" x 23" strips, lavender

Also:
- Cut 2: 3/4" x 5 3/4" strips, green print, for the Daffodil stems

DIRECTIONS

Refer to the photo for placement of the appliqué pieces. NOTE: *Keep appliqué designs centered when trimming the blocks.*

Clouds block:

- Arrange pieces #1, #2, and #3 on a 6 1/2"
  mottled blue square, placing the straight
  edges 1/4" from the edges of the square.
- Remove #3 and pin and appliqué the
  remaining clouds in place.
- Pin and appliqué #3 in place.
- Trim the block to 6" square.

Nest block:

- Arrange #1 through #6 on a 6 1/2" mottled
  blue square. Remove all but #1. Pin and
  appliqué it in place. Keep the outer edges
  at least 3/4" from the raw edges of the
  square.
- Pin and appliqué #2 through #6 in
  numerical order.
- Trim the block to 6" square.

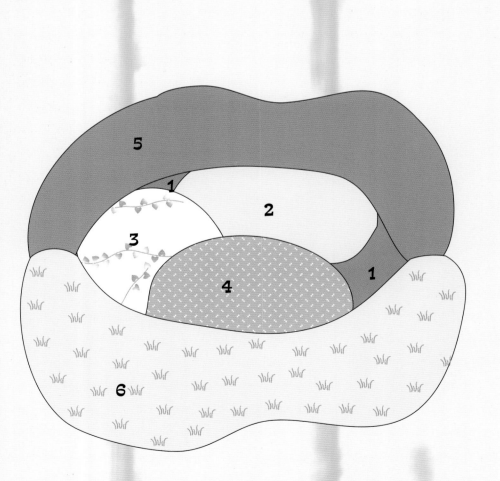

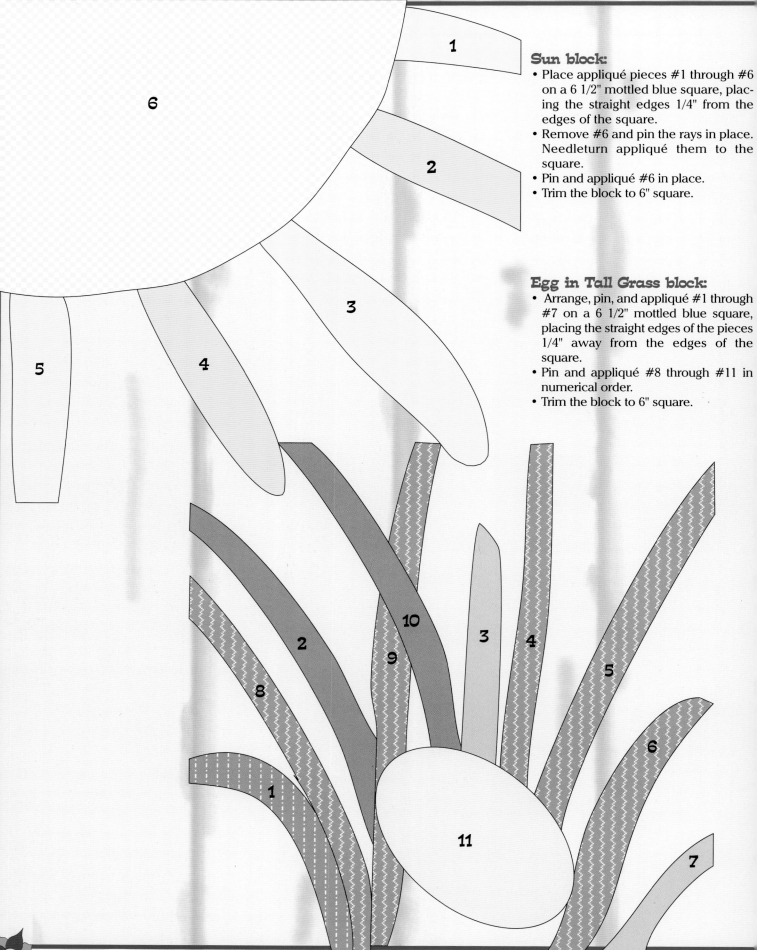

Sun block:
- Place appliqué pieces #1 through #6 on a 6 1/2" mottled blue square, placing the straight edges 1/4" from the edges of the square.
- Remove #6 and pin the rays in place. Needleturn appliqué them to the square.
- Pin and appliqué #6 in place.
- Trim the block to 6" square.

Egg in Tall Grass block:
- Arrange, pin, and appliqué #1 through #7 on a 6 1/2" mottled blue square, placing the straight edges of the pieces 1/4" away from the edges of the square.
- Pin and appliqué #8 through #11 in numerical order.
- Trim the block to 6" square.

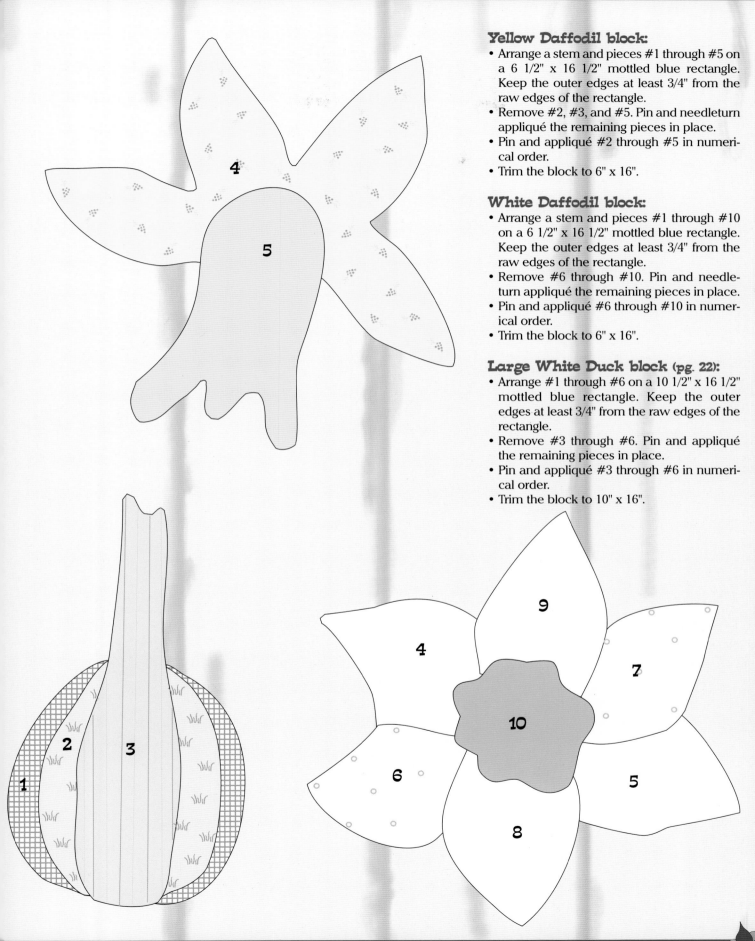

Yellow Daffodil block:

- Arrange a stem and pieces #1 through #5 on a 6 1/2" x 16 1/2" mottled blue rectangle. Keep the outer edges at least 3/4" from the raw edges of the rectangle.
- Remove #2, #3, and #5. Pin and needleturn appliqué the remaining pieces in place.
- Pin and appliqué #2 through #5 in numerical order.
- Trim the block to 6" x 16".

White Daffodil block:

- Arrange a stem and pieces #1 through #10 on a 6 1/2" x 16 1/2" mottled blue rectangle. Keep the outer edges at least 3/4" from the raw edges of the rectangle.
- Remove #6 through #10. Pin and needleturn appliqué the remaining pieces in place.
- Pin and appliqué #6 through #10 in numerical order.
- Trim the block to 6" x 16".

Large White Duck block (pg. 22):

- Arrange #1 through #6 on a 10 1/2" x 16 1/2" mottled blue rectangle. Keep the outer edges at least 3/4" from the raw edges of the rectangle.
- Remove #3 through #6. Pin and appliqué the remaining pieces in place.
- Pin and appliqué #3 through #6 in numerical order.
- Trim the block to 10" x 16".

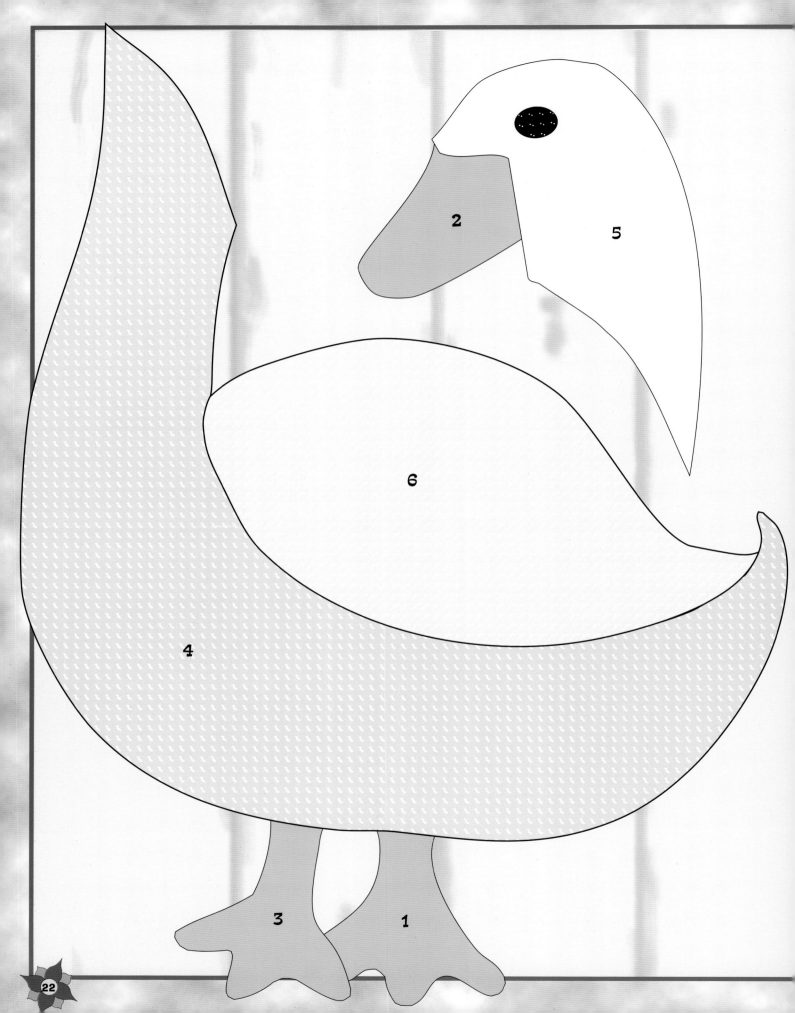

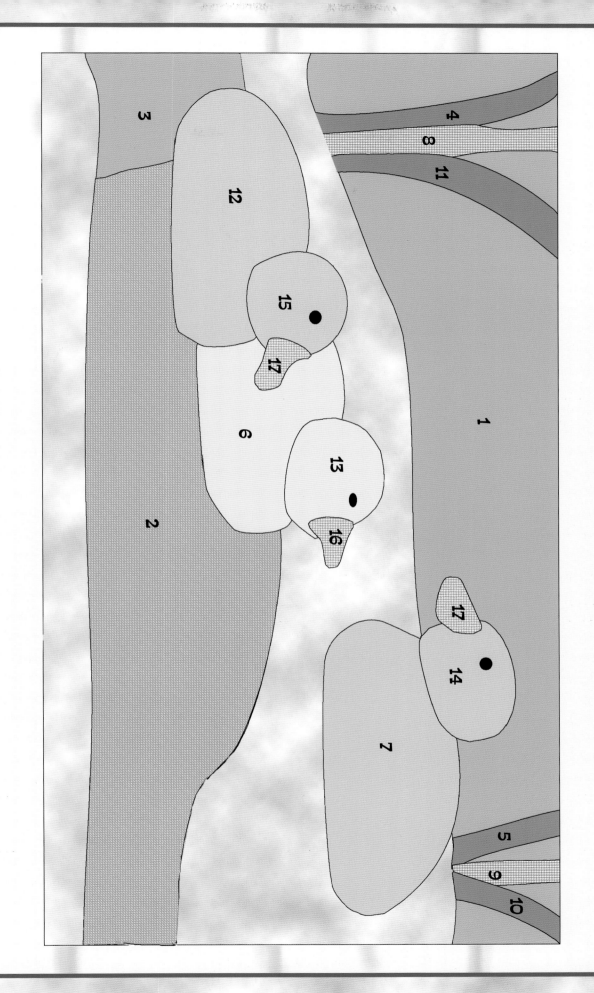

Duckling block (pg. 23):

- Arrange #1 through #3 on a 6 1/2" x 10 1/2" mottled blue rectangle, placing the straight edge of the pieces 1/4" from the edges of the square.
- Remove #3 and pin and appliqué #1 and #2 in place.
- Pin and appliqué #3 through #17 in numerical order.
- Trim the block to 6" x 10".

Bird on Branch block:

- Arrange pieces #1, #2, and #3 on a 6 1/2" x 10 1/2" mottled blue rectangle, placing the straight edges of pieces #1 and #2 one-quarter inch from the edges of the square.
- Remove #3 and pin #1 and #2 in place. Needleturn appliqué them to the rectangle.
- Pin and appliqué #3 in place.
- Pin and appliqué the remaining pieces in numerical order.
- Trim the block to 6" x 10".

ASSEMBLY

- Referring to the photo (pages 18), lay out the blocks in 3 rows of 3. Place 1 1/2" x 6" yellow strips between the smaller blocks and 1 1/2" x 16" yellow strips between the larger blocks.
- Stitch the blocks and strips into rows.
- Place the two 1 1/2" x 23" lavender strips between the rows. Join the strips and rows.
- Finish the quilt according to *Stitching Tips,* using the 1 1/4" x 44" yellow strips, for the binding.

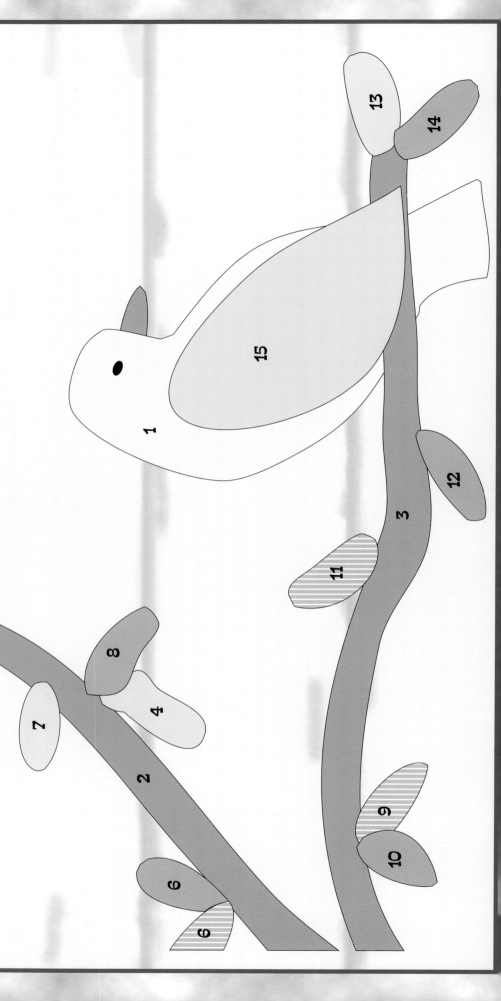

Dutch Partridge

Make your own partridge in a Dutch setting.

Color is the focal point of these delightful Dutch designs. **"Dutch Partridge"** is the perfect design to brighten any wall. Many wool projects include earth tones so the light colors of **"Dutch Partridge Tablemat"** make it a great stepping stone to new ideas for wool appliqué.

Materials for Quilt
FINISHED SIZE: 17" x 19"
- 5/8 yard pink print
- 1/4 yard teal print
- Fat quarter (18" x 22") yellow print
- Fat quarter turquoise print
- Fat eighth (11" x 18") blue print
- Yellow print, at least 9" square
- Green check, green print, salmon pink, and gold print, each at least 6" square
- Peach print, at least 4" square
- Purple print and turquoise print, each at least 3" square
- Lilac print, at least 2" square
- 19" x 21" piece of backing fabric
- 19" x 21" piece of thin batting
- Black embroidery floss

Materials for Tablemat
(pg. 29)
FINISHED SIZE: 11" Oval
- Assorted wool scraps for the appliqué pieces, each at least 6" square
- 16" square, yellow wool
- 11" square, blue wool
- 8" square, dark green wool
- 7" square, rose wool
- 12" square, 100% cotton print, for the backing
- Wool embroidery thread, embroidery floss, or pearl cotton in colors to match or contrast with the wool pieces

CUTTING
Appliqué patterns (pages 26-28) are full size and do not include a turn-under allowance. Cut pieces #1 through #23 using appropriately colored fabrics. All other dimensions include a 1/4" seam allowance.
- Cut 1: 19" x 21" rectangle, pink print
- Cut 1: large oval, yellow print
- Cut 1: small oval, blue print
- Cut 12: #24 leaves, turquoise print
- Cut 12: #25 berries, salmon pink print
- Cut 12: #26 arcs, teal print
- Cut 3: 1 1/4" x 44" strips, teal print, for the binding

DIRECTIONS
- Position pieces #1 and #2 on the blue print oval, referring to the Placement Diagram (page 26). Pin and needleturn appliqué them in place.
- Working in numerical order and referring to the photo for placement, needleturn appliqué pieces #3 through #23 on the blue print oval.
- Center and appliqué the blue print oval onto the yellow print oval.
- Referring to the photo, pin and appliqué the #24 leaves and the #25 berries in place around the yellow print oval.
- Center and appliqué the yellow print oval onto the 19" x 21" pink print background.
- Referring to the photo, pin and appliqué the #26 teal arcs around the outside of the yellow print oval leaving a 1/2" space between the oval and the arcs, making sure the edges of the arcs touch each other.
- Embroider a French knot (page 28) for the bird's eye.
- Finish the quilt according to *Stitching Tips (page 31),* using the 1 1/4" x 44" teal print strips for the binding.

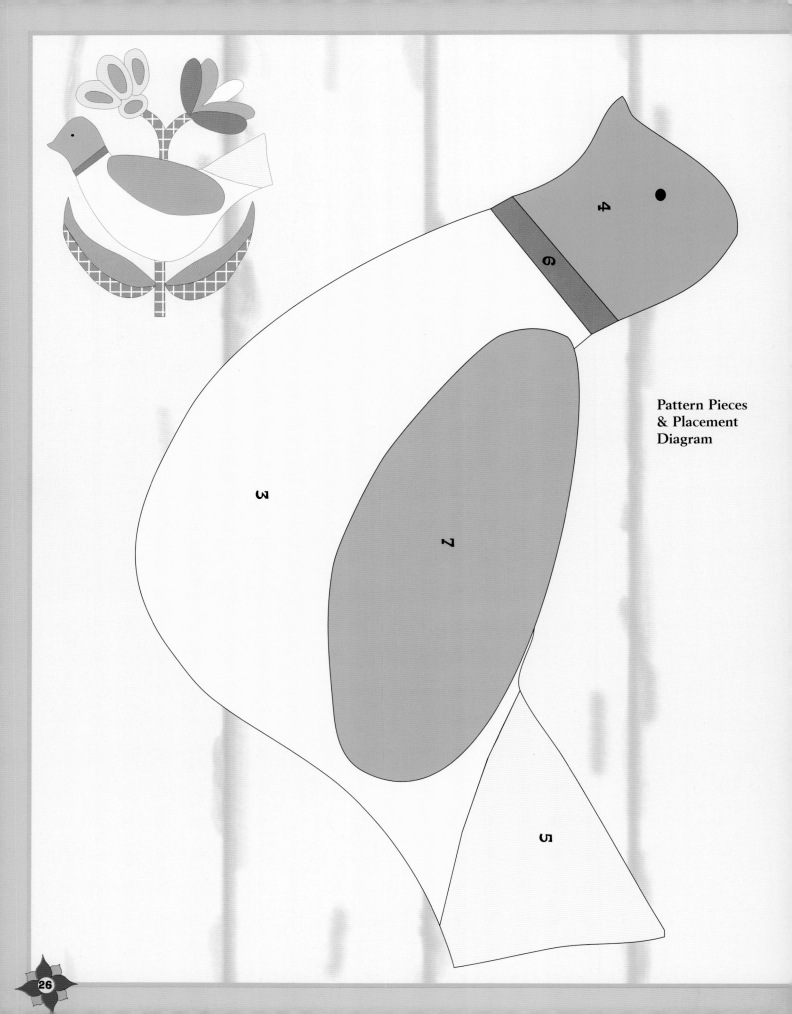

Pattern Pieces
& Placement
Diagram

4

6

3

7

5

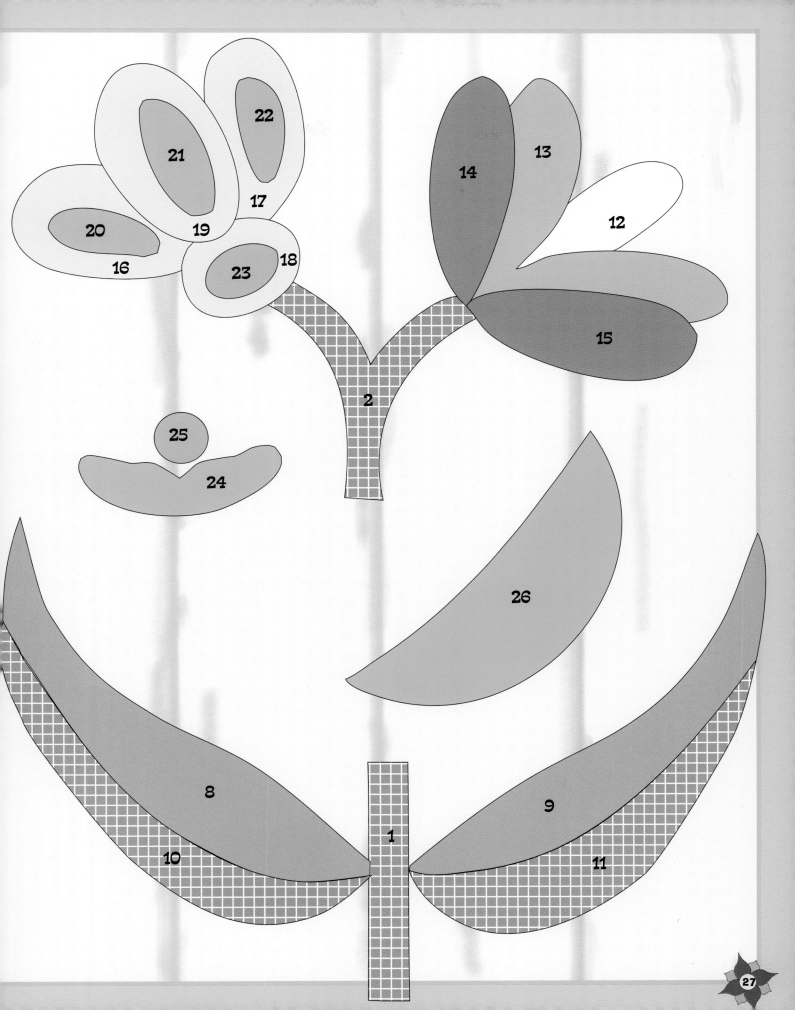

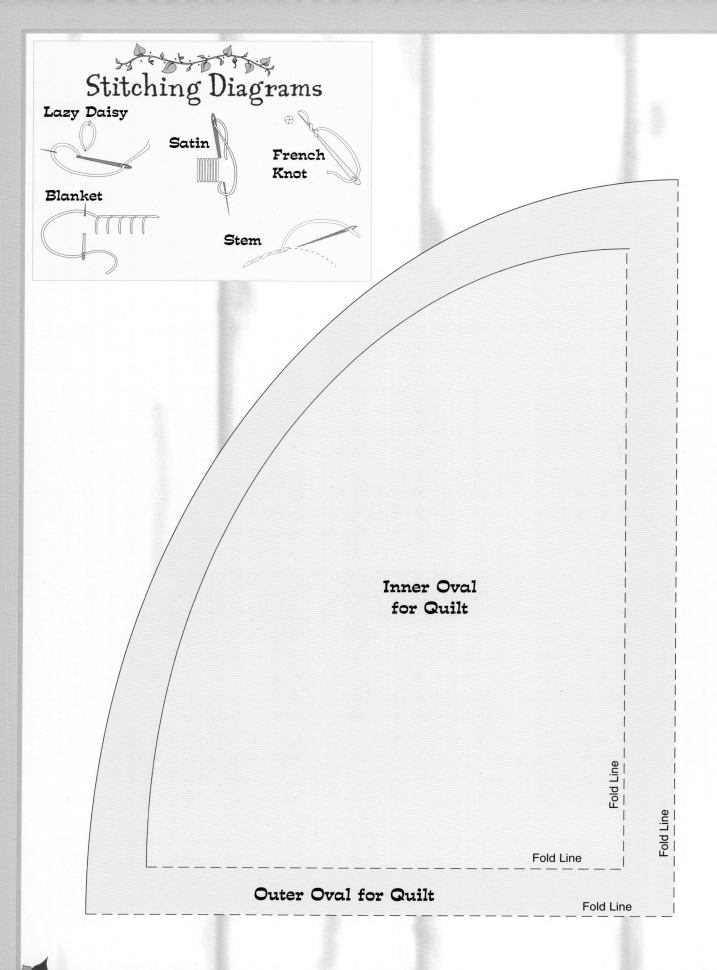

Stitching Diagrams

Lazy Daisy

Satin

French Knot

Blanket

Stem

Inner Oval
for Quilt

Fold Line

Fold Line

Fold Line

Outer Oval for Quilt

Fold Line

Dutch Partridge Tablemat

CUTTING

Appliqué patterns (page 30) are full size and do not include a turn-under allowance. Cut pieces #1 through #23 using appropriately colored wool.

- Cut 1: large oval, yellow wool
- Cut 1: small oval, blue wool
- Cut 1: large oval, backing fabric
- Cut 12: #24 leaves, green
- Cut 12: #25 berries, rose

DIRECTIONS

- Position pieces #1 and #2 on the blue oval, referring to the Placement Diagram. Pin and blanket stitch them in place. Refer to the opposite page for a stitch diagram.
- Working in numerical order and referring to the photo (page 25) for placement, pin and blanket stitch pieces #3 through #23 on the blue oval.
- Center and blanket stitch the blue oval onto the yellow oval.
- Referring to the photo, blanket stitch the #24 leaves and the #25 berries in place on the yellow oval.
- Embroider a French knot for the bird's eye. Refer to page 28 for a stitch diagram.
- Center the backing fabric on the back of the tablemat, wrong sides together.
- Blindstitch the backing fabric to the mat, turning under 1/8" to 3/16" on the backing fabric as you sew.

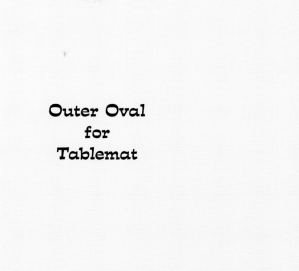

Fold Line

Fold Line

**Outer Oval
for
Tablemat**

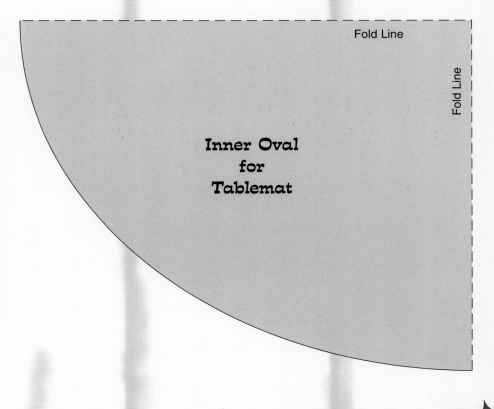

Fold Line

Fold Line

**Inner Oval
for
Tablemat**

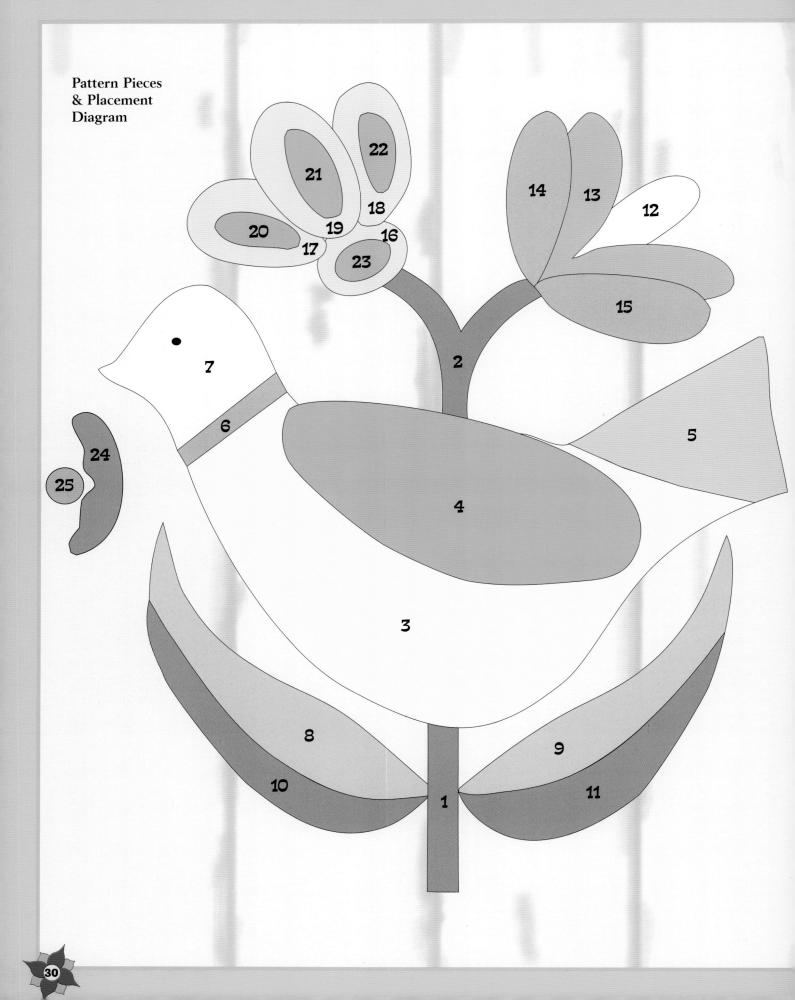

Pattern Pieces
& Placement
Diagram

Stitching Tips

ABOUT THE PATTERNS

Read through the pattern directions before cutting fabric for your project. Pattern directions are given in step-by-step order. Unless otherwise noted, all dimensions include a 1/4" seam allowance.

FABRICS
Cotton:

I suggest using 100% cotton. Wash fabrics in warm water with mild detergent and no fabric softener. Wash darks separately, and check for bleeding during the rinse cycle. Dry fabric on a warm-to-hot setting to shrink it. Press with a hot dry iron to remove any wrinkles. Yardage requirements are based on 44" fabric with a useable width of 42".

Wool:

I recommend 100% wool. Purchase felted wool when available, or felt your own in the following way: Wash wool fabric with detergent in very hot water. Dry it in the dryer on a hot setting. Allow for 10% shrinkage when felting. Yardage requirements are based on 60" fabric with a useable width of 58".

ROTARY CUTTING COTTON FABRIC

Begin by folding the fabric in half, selvage to selvage. Make sure the selvages are even and the folded edge is smooth. Fold the fabric in half again, bringing the fold and the selvages together, again making sure everything is smooth and flat.

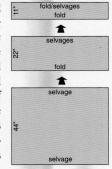

Position the folded fabric on a cutting mat so that the fabric extends to the right for right-handed people, or to the left for left-handed people.

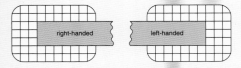

Place a ruler on the fabric, aligning a horizontal line with a folded edge. The ruler must be absolutely perpendicular to the folded edge. Trim the uneven edge with a rotary cutter. Make a clean cut

through the fabric, beginning in front of the folds and cutting through to the opposite edge with one clean (not short and choppy) stroke. Always cut away from yourself — never toward yourself!

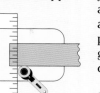

Move the ruler to the proper width for cutting the first strip, and cut.

Continue cutting until you have the required number of strips. Open up the fabric strip and check the spots where there were folds. If the fabric was not evenly lined up or the ruler was incorrectly positioned, there will be a bend at each of the folds in the fabric. To keep the cut edges even, always move the ruler, not the fabric.

When cutting many strips, check after every four or five strips to make sure the strips are straight. Leave the other strips folded in fourths until you are ready to use them.

PIECING

For machine piecing, set the machine at 12 stitches per inch, and stitch exactly 1/4" from the edge of the fabric. If desired, mark the throat plate with a piece of masking tape or moleskin 1/4" away from the point where the needle pierces the fabric. Align fabric edges with the tape or moleskin when stitching. Start and stop stitching at the cut edges unless otherwise noted. Backstitching is not necessary unless specified in the pattern, however, it is a good idea to backstitch when adding the outer borders to your quilt.

For hand piecing, begin with a small backstitch. Continue with a small running stitch, backstitching every 3-4 stitches. Stitch directly on the marked line from point to point, not edge to edge.

APPLIQUÉ
Needleturn

Appliqué pattern pieces are full size. Trace each pattern piece on the dull side of freezer paper and number the pieces as indicated. Cut on the drawn lines. Freezer paper templates may be reused numerous times, recut as necessary. Iron the freezer paper templates to the right side of selected fabrics. Use a pencil to lightly trace

along the edge of the templates. This is your stitching line. Cut out the shapes adding a 1/8" to 3/16" turn-under allowance around each piece. Remove the freezer paper. Pin the appliqué pieces to the background square in the order given in the directions.

Fusible Web

For larger appliqués, especially if they will have other appliqués fused on top of them, I trim the center from the cut fusible web piece before adhering it to the fabric. This reduces bulk and stiffness in the finished quilt. For example, for a flower that will have a center fused on top of it, trace the flower on the fusible web. Cut it out about 1/8" to 1/4" outside the line. Then cut 1/4" away from the line on the inside. Either discard the center or use it to make another appliqué piece.

Fuse the trimmed flower outline to the fabric, and cut the flower out on the line. Fuse the flower to the desired fabric according to the manufacturer's directions.

PRESSING

Press with a dry iron. Press seam allowances toward the darker of the two pieces whenever possible. Otherwise, trim away 1/16" from the darker seam allowance to prevent it from showing through. Press all blocks, sashings, and borders before assembling the quilt top. Press appliqué blocks from the wrong side, on a towel, to prevent a flat, shiny look.

FINISHING YOUR QUILT:
MARKING QUILTING LINES

Mark the quilt top before basting it together with the batting and backing. Chalk pencils show well on dark fabrics; otherwise use a very hard (#3 or #4) pencil or other marker for this purpose. Test your marker for removability first. Transfer paper designs by placing fabric over the design and tracing. A light box may be necessary for darker fabrics. Precut plastic stencils that fit the area you wish to quilt may be placed on top of the quilt and traced. Use a ruler to mark straight, even grids.

Outline quilting does not require marking. Simply eyeball 1/4" from the seam or stitch "in the ditch" next to the seam or the neighboring patch. To prevent uneven stitching, try to avoid quilting through seam allowances whenever possible.

Masking tape can also be used to mark straight lines. Temporary quilting stencils can be made from adhesive-backed paper or freezer paper and reused many times. To avoid residue, do not leave tape or adhesive-backed paper on the quilt overnight.

BASTING

Cut the batting and backing at least 4" larger than the quilt top. Tape the backing, wrong side up, on a flat surface to anchor it. Smooth the batting on top, followed by the quilt top, right side up. Baste the three layers together to form a quilt sandwich. Begin at the center and baste horizontally, then vertically. Add more lines of basting approximately every 6" until the entire top is secured.

QUILTING

Quilting is done with a short, strong needle called a "Between." The lower the number (size) of the needle, the larger it is. Begin with an 8 or 9 and progress to a size 10 or 12 as you become more experienced. Use a thimble on the middle finger of the hand that pushes the needle. Begin quilting at the center of the quilt and work outward to keep the tension even and quilting smooth.

Using an 18" length of quilting thread knotted at one end, insert the needle through the quilt top and batting only, and bring it up exactly where you will begin. Pop the knot through the fabric to bury it in the batting. Push the needle with the thimbled finger of the upper hand and slightly depress the fabric in front of the needle with the thumb. Redirect the needle back to the top of the quilt using the middle or index finger of the hand underneath the quilt.

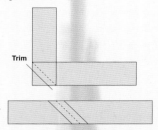

Repeat for each stitch, using a rocking motion. Finish by knotting the thread close to the surface and popping the knot through the fabric to bury it in the batting layer. Remove basting when all the quilting is done.

If you wish to machine quilt, I recommend consulting one of the many excellent books available on that subject.

If you intend to send your quilt top to a professional machine quilter, consult them concerning the necessary size of the batting and backing fabric.

SINGLE-FOLD BINDING

For most straight-edged miniature quilts, a single-fold binding is an attractive, durable and easy finish.

Stitch the binding strips together with diagonal seams; trim and press the seams open.

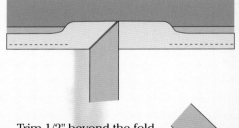

Trim one end of the strip at a 45° angle. Press one long edge of the binding strip 1/4" toward the wrong side. Starting with the trimmed end, position the binding strip, right sides together, on the quilt top, aligning the raw edge of the binding with the bottom edge of the quilt top. Leaving approximately 2" of the binding strip free, and beginning at least 3 inches from one corner, stitch the binding to the bottom of the quilt with a 1/4" seam allowance, measuring from the edge of the binding and the quilt top.

When you reach a corner, stop the stitching line exactly 1/4" from the edge of the quilt top. Backstitch, clip the threads, and remove the quilt from the machine. Fold the binding up and away from the quilt, forming a 45° angle, as shown.

Fold the binding down as shown, and begin stitching at the edge.

Continue stitching around the quilt in this manner to within 2" of the

starting point. Lay the binding flat against the quilt, overlapping the beginning end. Open the pressed edge on each end and fold the end of the binding at a 45° angle against the angle on the beginning end of the binding. Finger press the fold.

Trim 1/2" beyond the fold line. Place the ends of the binding right sides together, and stitch with a 1/4" seam allowance.

Finger press the seam allowance open.

Place the binding flat against the quilt, and finish stitching it to the quilt. Trim the batting and backing even with the edge of the quilt top. Fold the binding over the edge of the quilt, and blindstitch the folded edge to the back, covering the seamline.

DOUBLE-FOLD BINDING

Stitch the 2 1/2"-wide strips together with diagonal seams. Fold the strip in half lengthwise, wrong side in and press. Trim one long end of the strip at a 45° angle. Starting with the trimmed end, position the binding on the quilt top aligning the raw edge of the binding with the raw edge of the quilt top. Leaving approximately 6" of the binding strip free, and beginning at least 6' from one corner, stitch the binding to the quilt as described under Single-Fold Binding, stopping approximately 6" from the starting point. Finish the binding as described under Single-Fold Binding.